Daniel Hack Tuke

Reform in the Treatment of the Insane

early history of the retreat, York - its objects and influence, with a report

of the celebrations of its centenary

Daniel Hack Tuke

Reform in the Treatment of the Insane

early history of the retreat, York - its objects and influence, with a report of the celebrations of its centenary

ISBN/EAN: 9783337381325

Printed in Europe, USA, Canada, Australia, Japan

Cover: Foto ©berggeist007 / pixelio.de

More available books at **www.hansebooks.com**

REFORM IN THE TREATMENT OF THE INSANE.

EARLY HISTORY OF THE RETREAT,

YORK;

ITS OBJECTS AND INFLUENCE,

REPORT OF THE CELEBRATIONS OF ITS CENTENARY

BY

D. HACK TUKE, M.D., LL.D.,

Formerly Visiting Physician to the Retreat.

LONDON:
J. & A. CHURCHILL,
11, NEW BURLINGTON STREET.
1892.

To the Memory of

WILLIAM TUKE,

Whose Courageous Humanity a Century Ago

Is Recognized at Home and Abroad,

This Sketch is Dedicated by his Great-grandson,

THE AUTHOR.

"All men seem to desert me in matters essential."—*W. T.*

"Kind and conciliating treatment is the best means to promote recovery, as proved in the management of the Retreat, where coercion, though sometimes necessary for feeding the patients and preserving them from injury to themselves or others, is administered in the most gentle manner, and the use of chains is never resorted to."—*William Tuke.*

"The York Asylum has been wrested from its original design; the poor are in a great measure excluded, and the Institution, it is understood, is committed to the care of a physician and apothecary, without the interference of any committee or visitors in the internal management. Thus, instead of being a public charity, it has become a source of private emolument, and *hinc illæ lachrymæ.*"—*Henry Tuke.*

"If the 'Description of the Retreat' should be thought to afford satisfactory evidence in favour of a more mild system of treatment than has been generally adopted; if it should also prove, which I flatter myself it will, the practicability of introducing such a system into establishments for the insane poor, whose situation has, in general, been too pitiable for words to describe, I shall esteem myself peculiarly happy in this publication."—*Samuel Tuke.*

PREFACE.

SHOULD this sketch be deemed a dry narration of facts, the writer would plead in excuse that he has purposely restricted himself to them, in order to ensure historical accuracy, rather than allow sentiment and imagination to run wild in rhetorical reflections. In truth, the facts themselves ought to speak eloquently enough to those who have ears to hear. If, however, he has resisted the temptation to indulge in romance, and has rigorously confined himself to the records of the period which he describes, he is not the less impressed with the moral grandeur of the bold step taken a century ago in the interests of the insane. It was, indeed, a death-struggle between cruelty or neglect on the one hand, and kindness and consideration on the other. For long the issue trembled in the balance, but at last victory crowned unceasing labour. The imperative task and the great love, which Victor Hugo says, render men invincible, were not wanting.

Of those now living, comparatively few have the faintest conception of either the nature or the intensity of that struggle, the desperate efforts made to escape exposure and evade surrender, and the brilliant onslaught made upon long-established abuses. The languid interest now felt in the stirring events of that period (1792-1815) by the general public, and even many medical men, is not a little surprising, seeing that one in every three hundred of the population suffers from mental disorder and has good reason to be thankful that he is not lying in a dark cell on straw, " being bound in affliction and iron." His friends have also, one would have thought, some reason to be grateful. But, for all that, most of them would not care two straws for a narrative, compared with which a third-rate novel would excite more interest and emotion ; it is, in short, to the majority of people, a matter of profound indifference to know the cardinal facts of the history of the amelioration of the insane in England and France. At the annual dinner of the Medico-Psychological Association in 1881, Dr. Bucknill animadverted on the strange and discreditable contrast between the popular estimate of a victory like that achieved at the Retreat and a victory won on the field of battle.

There is a touching legend of a monk, who, as we have been lately reminded, wandered one day from his monastery into the adjacent forest, and, listening to its minstrelsy, did not return, oblivious of the flight of time, until fifty years had passed away. When he presented himself at the once familiar gates he was unknown, and he found, to his sad surprise, that scarcely one in the convent remembered his name. Had the principal actor in the scenes which were witnessed at York a hundred years ago revisited the Retreat this Midsummer, he would have found that his name, if not his person, was still remembered and reverenced. This would have been the reverse of the melancholy experience of the good monk, and he would surely have rejoiced, not, indeed, that his own name was held in remembrance by the company assembled on the scene of his labours, but that his strenuous and oftentimes painful endeavours had borne such remarkable fruit in his own country and abroad.

The writer embraces this occasion to thank, not only in his own name, but in that of the Retreat, the numerous friends of the insane in many lands, who, by their presence at the Centenary or by letter, rendered unstinted honour to the man who conceived the idea of the Retreat, and, thereby, of a new departure in the treatment of the insane, and whose dying words referred to the institution which, for nearly thirty years, he had served so faithfully and loved so well.

Lyndon Lodge, Hanwell, W.,
September, 1892.

REFORM IN THE TREATMENT OF THE INSANE.

EARLY HISTORY OF THE RETREAT, YORK;

ITS OBJECTS AND INFLUENCE.[*]

IN celebrating the Centenary of the York Retreat, the questions which arise in everybody's mind are, why was it established, and why at one time rather than at another? Further, it is natural to inquire what were its objects, and what influence has it exerted?

I. To answer the former questions we must briefly touch on the general condition of the insane a century, or rather more than a century ago, and also on the local circumstances which led up to the foundation of the institution. I shall not describe the dreadful suffering and neglect which existed in regard to the

[*] Paper read at the first Centennial Meeting of the Retreat, York, held at that Institution, May 6, 1892.

former condition of the insane. I may take it for granted that everyone here knows sufficiently well the deplorable state in which, for the most part, those labouring under mental afflictions were formerly to be found. He who doubts the truth of the descriptions given of the *bad* old times should visit the Guildhall Museum in London, and he will see there a specimen of the heavy chains formerly in use at Bethlem Hospital, and also the celebrated figures by Cibber of raving madness and melancholy, bound in fetters. The Treasurer and Governors of Bethlem have presented these relics of the past as the outward and visible sign of the blessed change which has taken place in asylum treatment. So far from being ashamed of them we glory in having them exhibited to the public eye, that the thousands of people who visit the Guildhall Museum may " Look here upon this picture, and on this."

It is interesting to refer for a moment to John Howard's incidental reference to asylums when, at the latter part of the 18th century, he was visiting prisons in various parts of the world. He says, " I greatly prefer the asylum at Constantinople to that of St. Luke's, or to Swift's Hospital at Dublin ;" but he appears to refer to the structure of the building, the rooms, the corridors, and the gardens, rather than to the condition of the patients themselves, for at Constantinople there was an asylum for *cats* near the Mosque of St. Sophia, where the *feline* inmates seem to have received more consideration than the human inmates of the asylum. Speaking of English prisons, in 1784 Howard observes that idiots and lunatics are confined in some gaols, and adds, " These serve for sport to idle visitors at assizes and other times of general resort. Many of the Bridewells are crowded and offensive, because the rooms which are designed for prisoners, are occupied by the insane." It is remarkable that more practical work was not done for the insane in England at this period when we remember the great interest which was excited in the disease by the fact that a distinguished Prime Minister, Lord Chatham, and the Sovereign himself had been laid low by mental disease. In fact, the attention of the nation had been concentrated upon the sick-room of George the Third and upon Dr. Willis, the clerico-medical doctor, who gained so much notoriety at that period. It was in April, 1789, that his Majesty went to St. Paul's to give thanks for his recovery, and enjoyed a lucid interval until 1801. It may be observed in passing that the treatment of the Royal patient was much on the lines of the prevalent doctrines of the day, perhaps not quite so depressing; and although there was nothing apparently to call for coercive methods, he was not only mechanically restrained by the doctor's orders, but was brutally knocked down by his keeper.

It would carry me quite too far, however interesting it might be, to recall what was happening in the world at the eventful period when this comparatively small work commenced at York. But what a contrast do the quiet proceedings which we commemorate to-day present to the wave of excitement which was passing over England as well as France, where the guillotine had just been invented and the King's fate was rapidly approaching. If we turn to the Annual Register of that period its pages are full of addresses from political societies in this country to the French National Convention. The preface to this volume asserted that "metaphysicians, geometers, and astronomers have applied the compasses of abstraction to human passions, propensities, and habits. The minds of men are alienated from kings and become enamoured of political philosophy." It may be said of some of the great events of this period that splendid and magnificent as they were when contrasted with lesser achievements, the world might have been better had they never occurred. Washington exclaims in one of his letters, " How pitiful in the eye of reason and religion is that false ambition which desolates the world with fire and sword for the purpose of conquest and fame, when compared with the minor virtues of making our neighbour and our fellow-men as happy as their frail condition and perishable natures will permit them to be."

And it was this very thing, " the minor virtue of making our fellow-men as happy as their frail condition permits them to be," that characterized the proceedings at York a hundred years ago. It was a time of local as well as national excitement, when the Corporation of York presented Charles James Fox with the freedom of the city in recognition of the efforts which he had made on behalf of liberty and the rights of man.

I must now ask your attention to the earlier year of 1777, when an asylum was opened at York in consequence of the need felt for such an institution for the insane poor in this locality. It was commenced under favourable auspices and evidently with the best intentions. It was not very long, however, before its management became unsatisfactory. I wonder how many people know that Mason, the Poet and Precentor of York Minster, was something more than either, and, in conjunction with Dr. Burgh and Mr. Withers, of York, endeavoured to hold the Governors of that day to the original design of the institution. They were persistently thwarted in their honourable endeavours. In 1788 Mason published his "Animadversions" on the asylum. In 1789 he was the means of procuring a legacy, which afterwards constituted "Lupton's Fund" for the poor, but from unworthy motives this charity was opposed by the physician of the asylum, and the Governors even passed a resolution in 1791 that anyone who contributed to it (and among those who did so was no less a person than Wilberforce) should be excluded from the privilege of being a Governor. In fact, in spite of these praiseworthy efforts, nothing whatever was done to remove abuses, and Jonathan Gray, the historian of the old York Asylum, wrote, " The opponents seem to have abandoned the matter as hopeless," and pathetically adds: " It cannot be doubted, therefore, that Mason, Burgh, and Withers quitted the world under an impression that their labours in this benevolent cause had been worse than useless, having been repaid only by obloquy and misrepresentation." Although, however, they were thus hopeless about the York Asylum, they rejoiced to know that an important step had been taken in establishing another institution. And this brings me to the well-known local incident which occurred in 1791.

A female patient was admitted from a considerable distance into the York Asylum. After a time her relatives desired and authorized some of their friends in York to visit her. They met with a repulse from the asylum authorities, and not long afterwards the patient died. That there are cases when a superintendent is fully warranted in advising the relatives and friends not to see a patient cannot be denied, and it would be, therefore, very unreasonable to ground serious complaint on the simple refusal of the superintendent to allow a lunatic patient to be visited. However, in this instance, as the patient was very ill and her friends were forbidden to see her, suspicions were naturally aroused, and further inquiries made after her death suggested neglect and possible cruelty. At this juncture William Tuke, a philanthropic citizen of York, was informed of the circumstances. He felt strongly that there was something wrong, not only in this case, but in the general management of the institution. He was not given to listen readily to sensational reports; his temperament was certainly sufficiently calm, and indeed his character, if contemporary descriptions are worth anything, was typical of enthusiasm without fanaticism, human sympathy without intrusiveness, philanthropy without fads. His portrait on the wall is expressive, I think, of this kind of man. The evidence, therefore, must have been of a very decisive character to induce him to arrive at this conclusion. He knew that any direct attack upon the asylum would meet with the same fate as that which disheartened Mason and Dr. Burgh; but his mind was stirred within him, and he began to think whether it would not be desirable and possible that an institution should be established in which, without destroying privacy, there should be no secrecy in its relation to the family of the patient, and in which the inmates should be treated with humanity. Thus revolving the subject in his mind, he arrived at the conclusion that the question ought to be answered in the affirmative. He conferred with his friends. Some of them took the same view as himself, especially his son and daughter-in-law, Henry and Mary Maria Tuke, who warmly supported the idea, as also did his excellent friend, Lindley Murray, the grammarian. His own wife, although she had been a helpmate in some of his benevolent schemes, did not favour this, and, being of a satirical turn of mind, said he had had many children emanate from his brain, and that " his last child was going to be an idiot." Who shall say how many of the great designs of men have been nipped in the bud by the ridicule of women ! However, he was not one to be easily discouraged either by opposition or by satire, and the result was that in the spring of 1792, he brought forward a definite proposition at the close of, and altogether distinct from, the business transacted at a quarterly meeting of the Society of Friends held at York, that an asylum for the insane should be established. No official record, therefore, was made of the conference. The proposition was thought to be one the wisdom of which admitted of grave doubt indeed ; a wet blanket was, in fact, thrown on the scheme, and the meeting broke up in this mood. Even several years afterwards we find him, on the brink of despair, writing a letter, in which he exclaims, " All men seem to desert me in matters essential." Many would no doubt have been permanently disheartened; but William Tuke made still further inquiry as to the necessity for such an institution with the effect of fortifying his position. He visited some of the asylums in repute at that period. At St. Luke's Hospital, London, he found a miserable state of things, chains, and a large number of patients lying, as he described them, naked and on filthy straw. His description recalls that given of Mrs. Fry's visit to an asylum at Amsterdam many years later, where she noticed but could not relieve an unhappy woman heavily ironed and similarly grovelling on the floor.

What this angel of mercy was unable to do at the Amsterdam asylum William Tuke was able to do at St. Luke's Hospital, so far as this, that a female patient who was thus chained to the wall and shamefully neglected was subsequently removed to the Retreat, and in one of his letters he speaks with gratification of the comfort thus afforded her.

Well, William Tuke, although he had received a check, returned to his charge and reinforced his arguments at a meeting held June 28th, 1892, three months after his first proposal in March. The opposition was renewed. One of those who were present on this stormy occasion has stated that the whole scheme seemed for some time as if it would be entirely shelved, so strong was the objection to it, but that the speech of Henry Tuke turned the scale, for if his father was the *fortiter in re*, the son was the *suaviter in modo*, which sometimes succeeds when the other alone fails. He said to the meeting, " Well, but isn't it worth while considering my father's proposition?" The consequence was that at this second meeting the Retreat was instituted, though not without the note of Cassandra being heard, and, therefore, assembling as we do in this month of May to celebrate it, we meet very appropriately at a time intermediate between the first proposition in the spring and its formal institution in the midsummer of 1792, and can vividly realize the anxiety which must have filled the breast of the projector as to whether his scheme would be crushed or accepted.

The opposition to the proposal is not surprising when we consider how little was known at that time of the condition of the insane, or of what might be done in the way of treatment and kindly moral management. I have already said that conspicuous among those who listened to the proposition was the well-known Lindley Murray, who not only gave what I may call his " Grammar of Assent " to the undertaking, but was helpful from time to time in giving that which was far better than money—his calm judgment and thoughtful advice as to the best mode of proceeding—employing just that diplomatic way of going about the business which succeeded in winning over objectors and lukewarm friends in support of the experiment, which Lindley Murray so well knew how to employ on critical occasions—a man so justly respected for his worth, his kindly nature, and the judicial character of his mind. To most he is known only as the Grammarian, and I suppose there are many who wish they had never formed his acquaintance in this character when at school; but he ought to be remembered with respect for the wise counsel which he gave in connection with the early history of this institution.

A learned Professor of Chemistry in an American College, when travelling in Europe, visited the recluse at Holdgate at this time, and in his book, giving an account of his travels, he records this visit with great pleasure, and writes : " Who would not rather be Mr. Murray, confined to his sofa, than Napoleon, the guilty possessor of a usurped crown and the sanguinary oppressor of Europe?" I fear that, in this wicked world, all would not reply as the Professor anticipated!

When I was in America some years ago I was requested to be present at a social gathering in the institution at Northampton, Massachusetts, over which the veteran alienist, Pliny Earle presided, as medical superintendent, and in the speech which he made nothing was more interesting to the audience than the statement that when a young man he visited York and had the pleasure of finding in the bedroom which he occupied in Samuel Tuke's house, the wheeled chair which was used for many years by the Grammarian, who, as you know, met with an accident in his native land (America), in consequence of which he had only the partial use of his lower extremities.

As a consequence of the Resolution which was passed at the midsummer meeting, ground was purchased in a suitable and healthy locality near York, a city then of 16,000 inhabitants. The locality itself was historically interesting, for it contained a mound on which at that time stood a windmill, from which it is supposed that its name, "Lamel Hill," was derived, "being no more," says Drake, "than *le meul*, miln hill, called so by the Normans." Its height above the summer level of the Ouse was about 90 feet. Here it was that the troops of Fairfax and Lesley placed their battery during the siege of York by the Parliamentary Army in 1644, symbolical, we may say, of the fight made by those whose weapons were not carnal against the cruel treatment of the insane, while they laid siege to the whole system of asylum abuses. But we must hasten on and think rather of the new Institution itself, of which, in its original state, there is a representation here from a drawing taken by a York artist, Mr. Cave. The building bore no resemblance to the prison-like asylum of the day, and a special point was made of avoiding bars to the windows; but time will not allow of my entering into any details, important as I consider them to be. If these old windows now excite criticism, let it be remembered that at Bethlem Hospital, even in 1815, the bedroom windows were unglazed.[*]

On the foundation stone, which Macaulay's New Zealander may some day find among the ruins of the Retreat, were inscribed the words :—

Hoc Fecit
Amicorum Caritas in Humanitatis
Argumentum
Anno Dni MDCCXCII.

[*] "Report of the Select Committee of the House of Commons, 1815."

[Cave pinx.] [Cooper del.]

ORIGINAL BUILDING OF THE RETREAT, YORK. INSTITUTED 1792.

This inscription is of great interest and importance, as proving that in 1792 the word HUMANITY was uppermost in the minds of the friends of the movement— their leading idea. "The charity or love of friends executed this work in the cause of humanity." In other words, charity raised the edifice as a token or sign in demonstration of humanity. It is also interesting to note that on the foundation stone, not actually laid until later, the period of instituting the Retreat was carefully recorded as 1792, as, indeed, it was on the first page of the early Annual Reports.*

But in dwelling on the foundation stone we must not forget the important matter of the name which was given to the institution, and this, like the inscription, carried with it a deep meaning. I have said that the wife of William Tuke indulged in some sarcasm with

* When the Retreat was projected the great mass of the insane in England were unprovided for as regards asylum accommodation. In addition to three or four private asylums, including Ticehurst, there were the well-known, but unfortunately ill-managed hospitals of Bethlem and Saint Luke's, and the lunatic ward of Guy's Hospital. There were, at Manchester and Liverpool, wards for the insane in connection with the Royal Infirmaries of those towns, and in addition to the old York Asylum there was the Norwich Bethel Hospital, and St. Peter's Hospital at Bristol, to which, many years after, the celebrated Dr. Prichard was physician. The recognized number of insane in London and in the country was under 7,000, which stands out in strange contrast with the number registered at the present day.

regard to his proposal. It was very different, however, with his daughter-in-law, Mary Maria Tuke, and when the inevitable question arose and was discussed in the family circle, " What name shall we adopt for the new establishment ?" she quickly responded "The Retreat"—a name, be it remembered, which up to that time had never been applied to an asylum for the insane; in fact, in the vulgar tongue, an asylum was, as I have said, a madhouse—this and nothing more. Surely, it was a most felicitous term, and a beautiful illustration of that aspect of the movement uppermost in the minds of those who were engaged in the undertaking, that, as is stated, " It was intended to convey by this designation their idea of what such an establishment should be, namely, a place in which the unhappy might obtain a refuge; a quiet haven in which the shattered bark might find a means of reparation or of safety." I wish that I had the happy power of reviving or restoring the picture of the interior of the early Retreat life as I seem to see it myself. William Tuke's brother-in-law, T. Maud, a surgeon in Bradford, was to have helped him in carrying forward his plans and resided at the Retreat, but this arrangement was unfortunately cut short by his unexpected death, and William Tuke had to superintend it himself. My father, in his " Review of the Early History of the Retreat," writes : " The Founder looked around among his friends for a suitable successor, but not finding one ready for the engagement, he agreed to take the office himself till a substitute should be found ; and for nearly twelve months he had the immediate management of the young establishment upon him," and for about thirty years, having retained his paternal interest in it, inspired its proceedings. There was, then, William Tuke, the father of the little family, organizing, planning, and arranging the details of the house, and planting with his own hands the trees which we now see on the north boundary of these grounds. Then there was a physician at that time in York, Dr. Fowler, who, in this capacity, visited the Retreat, and was a kindly, estimable, and unassuming man. He is described as one " who estimated men and things according to their real value rather than their names or aspects." He originally came from Stafford. I am unable to obtain any particulars with regard to his life ; but his name is associated with what is called " Fowler's Solution," the well-known preparation of arsenic in use at the present day. He died, much regretted, five years after the opening of the institution, and was succeeded in the office of visiting physician by a young and ardent physician, Dr. Cappe, whose talents and affectionate disposition gave promise of a useful career, and who felt a warm interest in the Retreat. He threw his whole soul into his work, but grave pulmonary symptoms soon made their appearance. He sought in vain to recover his health in a warmer clime, and, to the sorrow of all connected with the Institution, fell a victim to consumption.

But to return. Patients were being admitted, and were kindly cared for and treated. I have mentioned the poor woman who was brought from St. Luke's, and I may add that there was another patient (a man), who, when admitted, was found to have lost the use of his limbs, and when released from his manacles tottered about like a little child, but regained the use of his muscles and required no mechanical restraint. When visited by one of his relatives and asked what he called the Retreat, he replied, with great warmth, "Eden, Eden, Eden!"

And now I must hasten to speak of one who was largely influential in carrying out the hopes and aims of the original projector. This was George Jepson, a most estimable man residing at Bradford. My father, who greatly appreciated him, writes:—"He was almost entirely a self-taught man; yet so highly esteemed in his neighbourhood, that he was the counsellor of many of the country people for miles around his residence, in some of their most important private concerns; and he may be said to have been a medical practitioner." He by no means confined himself to the medical art; in fact, he never passed any examination, for at that period it was not illegal to practice without a qualification. He was an acute observer, and one who thought for himself. It was in 1797 that he was induced to come to the Retreat. It certainly was not the amount of medical knowledge which he possessed, but rather freedom from the trammels of the medical schools of the day (although at first he had a prejudice in favour of the lancet), which rendered him a suitable person to be appointed to the Retreat. My father thus writes of this period: —"George Jepson was of course initiated into the duties of his office by William Tuke. It was indeed," he remarks, "a rare concurrence of circumstances which brought together two minds, one so capable to design largely and wisely, the other so admirably fitted to carry such designs into execution. The two men, though exceedingly different, were one in an earnest love to God and man, in disinterestedness and decision of character, and therefore in a steady constant perseverance which works outward wherever truth and duty lead." It may also be stated that when Sydney Smith visited the Retreat at a later period he was much struck with Jepson, and his wife also, who acted as matron. The Grand Duke Nicholas, afterwards the Emperor of Russia, on going round the Retreat, was impressed by her appearance, and remarked in a low tone to my father, " Quel visage ! " No man was more esteemed and beloved by the projector of this institution and by his family, and I am quite sure that if he could be with us to-day he would wish that due honour should be rendered to Jepson for what he did within these walls. There was but one feeling, that of mutual esteem. William Tuke rejoiced at being able to meet with a man who entered so readily into his schemes and acted so loyally in carrying them out ; while Jepson looked upon " the Manager-in-Chief " (as my father designates William Tuke) as his " guide, philosopher, and friend." It was William Tuke's custom to correspond with a medical nephew and to communicate to him what they were doing at the Retreat, especially as to the results of the then heterodox treatment pursued. My father attached great value to these letters, and I have a vivid recollection of the pain which he experienced in consequence of a number of them being carelessly destroyed by a domestic who, in her ignorance of their value, had torn them into shreds, and had been using them for her candles. In one of the letters which remain (and some are only fragments) written in 1798, and addressed from the Retreat, I find him discussing the value of opiates, although without the advantage of a medical education; while in other letters he refers with lively interest to the utility of the warm bath. Of course, all this was very wrong from my own professional point of view, but there was some excuse for it when we consider the state of mental medicine at that time in York. Why, the physician of the old York Asylum boasted of his " secret insane powders, green and grey," which, as Dr. Thurnam states, "were sold as nostrums for insanity throughout a great part of Yorkshire and the north of England."

In another letter, dated from the Retreat in that year, and before William Tuke had had the good fortune to meet with Jepson, he mentions the case of a female who, on the way to the institution, "dreaded being put into a kind of dungeon." When visited, the morning after her arrival, she promised him that if she might only stay at the Retreat she would behave well, and she requested her daughter who had accompanied her to return home. On this he makes the commentary, "A strong proof of the sensibility of insane persons respecting those who have the care over them." With delight he reported that he had almost every day observed an improvement in the case of a patient among those first admitted who had occasioned him great anxiety. In one instance a patient committed suicide, and he was greatly distressed. He relieved his mind in a letter to his friend, the well-known philanthropist Richard Reynolds, and received a very sympathetic letter in reply.

I may remark that it is not very difficult to understand the successful treatment of the patients at the Retreat, although there may have been little of that definite scientific or medical element which is so justly prized at the present day. But although there was not over much science and still less medicine in the primeval atmosphere of the Retreat, the single-mindedness of those who were trying what may be called a Holy Experiment—that of personal kindness and love to man in his misfortune and sickness as well as in health—helped to secure its success. It must not be supposed that medicine was despised. It is true that a clean sweep was made of the routine of bleeding, blistering, purgatives and emetics then in vogue in what were regarded as the best institutions for the treatment of the insane, and this probably gave rise to the idea that Æsculapius was not duly honoured at the Retreat; but there was the guarded use of drugs, a careful attention to the general health, and a very special use of the warm bath. It was also found that instead of lowering the patient it was generally better to feed him, and that good nights could be obtained for the excited, not by antimony and other depressants, but by good malt liquor. Medical men were scandalized at such a reverse in the mode of treating excitement and sleeplessness, but it was acknowledged before long that the results were of the happiest kind.

And here we shall be assisted in forming an idea of the management and treatment pursued at the

Retreat by the evidence which William Tuke gave before the Select Committee of the House of Commons, although of course he did this at a much later period. The new system had become widely known, the old system was on its trial, and the Parliamentary Committee naturally called upon the projector of the Retreat to supply them with information in regard to its management and the treatment pursued there. I have heard my father, who accompanied him, speak of the great interest which his presence excited. The witness spoke with pleasure and satisfaction of what had been effected at the Retreat. After stating (in reply to a question) that he had taken an active part in everything that had been done respecting the institution from the beginning, he was asked to give to the Committee an account of the practice pursued in the establishment. He replied in general terms that "everything is done to make the patients as comfortable as they can be, and to endeavour to impress upon their minds the idea that they will be kindly treated; that is generally the setting out; when that is done it is not so difficult to manage the patients." Asked in regard to the effect of medicines in cases of mental derangement, he replied that he thought that very little could be done except when the disorder is accompanied by bodily disease of one kind or other. He said that from his personal observation he considered that patients had frequently recovered in consequence of the removal of the physical complaint. He was requested to inform the Committee whether the patients were periodically physicked, bled, made to vomit, and so forth, and he replied with great emphasis *"No such thing,"* and added, "That with respect to bathing the bath was frequently used, the warm bath more than the cold, but that in no case was it employed periodically. It was his opinion that the warm bath had been found very beneficial." The subject of mechanical restraint has become such a burning question in these latter days that it is interesting to ascertain from his evidence what was the actual practice at the Retreat. It has often been stated in histories of the treatment of the insane in England that the Retreat introduced what is called non-restraint. This is quite a mistake. It never was and is not at the present day a dogma held by those who have the management of the Retreat that under no circumstances whatever is it justifiable to resort to mechanical means of restraint.

On the contrary, it was frequently stated by those who spoke in the name of the institution that no rule could be laid down on the subject, and that it must be left entirely to the discretion of the medical superintendent so long as he retains the confidence of the Directors. William Tuke stated to the House of Commons Committee that in violent cases it was found necessary to employ, sometimes, a leather belt to confine the arms, and that this was preferred to the strait-waistcoat on account of its not heating the body so much, and leaving the hands free for use, although not so much as to do mischief. Seclusion was resorted to, he said, when it was found necessary. Thus he says, "We have a patient who has long lucid intervals of calmness, but is subject to very violent paroxysms and very sudden ones, during which we conceive he would injure any person who came within his power; this man during his paroxysms is confined in a separate room, about 12 feet by 8." In this instance it seems that the strait-waistcoat was occasionally used, and William Tuke found it necessary to state that he did not permit the use of chains of any kind.

I hope that this evidence, along with the letters I have quoted, few as they unfortunately are, will convey a clear idea of the early, as well as the somewhat later, Retreat treatment of patients. I must for a moment retrace my steps to remark that one of the best proofs of the important work carried on in the early days of the Retreat was the striking impression produced upon visitors, especially medical men. Only two years after its opening a Swiss physician, Dr. de la Rive, bent his steps thither, was delighted with what he saw, and published a very interesting account of his visit. " This house," he wrote, " is situated a mile from York, in the midst of a fertile and cheerful country; it presents not the idea of a prison, but rather that of a large rural farm. It is surrounded by a garden. There is no bar or grating to the windows." In 1812 Dr. Duncan, of Edinburgh, who was greatly interested in the lunatic asylum of that city, also visited the Retreat, and spoke in the highest terms of its management. He considered that it had " demonstrated beyond contradiction the very great advantage resulting from a mode of treatment in cases of insanity, much more mild than was before introduced into almost any lunatic asylum either at home or abroad." He regarded it as " an example claiming the imitation and deserving the thanks of every sect and every nation. For, without much hazard of contradiction from those acquainted with the subject, it may be asserted that the Retreat at York is at this moment the best regulated establishment in Europe, either for the recovery of the insane, or for their comfort where they are in an incurable state."

When in Paris many years ago, I visited M. Ferrus, the first Napoleon's physician, and a distinguished alienist. He recalled in graphic terms and with that gesture-language in which the French so much excel us poor phlegmatic Englishmen, the pleasure and surprise he had experienced on visiting the Retreat. I subsequently found a description of his visit in print. There he refers to it as "the first asylum in England which attracted the notice of foreigners;" and describes its projector as "a man for whom religion and morality were practical virtues, and in whose eyes neither riches, nor poverty, nor imbecility, nor genius ought in the slightest degree to affect the bonds which unite all men together in common. He thought with reason that justice and power ought to be evinced, not by shouts and menaces, but by gentleness of character and calmness of mind, in order that the influence of these qualities might make themselves felt upon all, even when excited by anger, intoxication, or madness.

The traditions of this friend of humanity are preserved in the house which he founded." M. Ferrus adds that " those who are admitted find repose in this building, which much more resembles a Convent of Trappists than a madhouse; and if one's heart is saddened at the sight of this terrible malady, one experiences emotions of pleasure in witnessing all that an ingenious benevolence has been able to devise to cure or alleviate it."

A pleasing picture of the interior of the Retreat is given in a poem written more than 80 years ago. Many here are no doubt familiar with certain lines of Wordsworth, headed " To the spade of a friend, an agriculturist, composed while we were labouring together in his pleasure ground." His friend's name was Wilkinson, a minor Lake poet, who, on visiting the Retreat 14 years after it was opened, described it in verse too long to cite here, but from which I may take the following few lines :—

> " On a fair hill, where York in prospect lies,
> Her towers and steeples pointing to the skies,
> A goodly structure rears its modest head ;
> Thither, my walk the worthy Founder led.
> Thither with Tuke, my willing footsteps prest,
> Who oft the subject pondering in his breast,
> Went forth alone and weigh'd the growing plan,
> Big with the lasting help for suffering man."

I must not occupy your time in quoting more from this poem than the lines which bring before us in a vivid manner the social and homely character of the group of patients whom he describes, and which appears to have removed from his mind the apprehensions with which he entered " The Wards of Insanity," as he calls them :—

> "Such and so on I passed with fearful tread,
> With apprehensive eye, and heart of lead ;
> But soon to me a motley band appears,
> Whose blended sound my faltering spirit cheers ;
> What female form but brightens into glee
> Whilst bending o'er exhilarating tea ?
> What man but feels his own importance rise,
> Whilst from his pipe the curling vapour flies ?
> But oft, alas ! tea and tobacco fail
> When demons wild the erratic brain assail.
> But why this wreck of intellect ? Ah ! why
> Does Reason's noble pile in ruins lie ?"

Whether Wilkinson's poetry is equal to that of his friend's " Excursion " I will not decide, but we cannot help feeling grateful to him for having left on record the impression produced upon his mind by the Retreat not many years after it was opened.

II. Now, what were the primary objects in view in the foundation of this Institution ?

First, the revulsion from the inhumanity which had

come to light rendered it necessary that the fundamental principles of moral treatment should be those of kindness and consideration for the patients. They were the basis of the proceedings which were taken; in fact, as we have seen, they were carved upon the very foundation stone of the building.

A *second* object undoubtedly was to provide an atmosphere of religious sentiment and moral feeling congenial to the accustomed habits and principles of those for whom the institution was primarily intended.

Thirdly, it is a significant fact that when the Retreat was instituted, it was laid down that there should be "a few acres for keeping cows and for garden ground for the family, which will afford scope for the patients to take exercise when that may be considered prudent and suitable." Recreation and employment were put prominently forward directly the Institution was opened, and were carried out into practice much to the surprise of those who visited the house. The Swiss physician (Dr. de la Rive), who in 1798 visited the Retreat, as I have related already, reported thus :—"As soon as the patients are well enough to be employed, they endeavour to make them work. The women are employed in the usual female occupations ; the men are engaged in straw and basket work, etc. The Institution is surrounded by some acres of land which belong to it. The superintendent had undertaken to make the patients cultivate this land, giving each a task proportioned to his strength. He found that they were fond of this exercise, and that they were much better after a day spent in this work than when they had remained in the house, or even when they had taken a walk."

Fourthly, the moral treatment must no doubt be emphasized as characteristic of the early practice of the Retreat. The physician just mentioned writes :— "You see that in the moral treatment they do not consider the insane as absolutely deprived of reason, that is to say, as inaccessible to the motives of hope, feeling, and honour ; rather they are regarded, it would seem, as children who have an excess of force and who make a dangerous employment of it." In the first Annual Report (written by W. Tuke) occurs the following:—"They who think the object worthy of their attention may be encouraged to promote it, not only on the principle of charity to the poor, but even from compassion to those in easy and affluent circumstances, who will, doubtless, think themselves benefited, though they may pay amply for it." It is pointed out that " those who have embarked in this undertaking have not been influenced by interested views, nor are they requesting or desiring any favours for themselves. A malady, in many instances, the most deplorable that human nature is subject to, hath excited their sympathy and attention." Lastly, an appeal is made for " co-operation in an Establishment which hath for its object the mitigation of human misery, and the restoration of those who are lost to civil and religious society, in the prosecution of which they humbly rely on the favour of Him whose tender mercies are over all His works." I may add that the title page of this Report bore the words : *" The State of an Institution near York, called the Retreat, for persons afflicted with Disorders of the Mind;"* certainly a very sufficient description of the object for which it was established, and this title page remained undisturbed until 1869, when, unhappily, as I think, it was discarded for another.

Fifthly, that which from the first has been regarded as a most important feature of the Institution, is its *homishness*—the desire to make it a family as much as under the peculiar circumstances of the case is possible

However desirable the scientific study of insanity may be, and I hope we shall never underrate it, it would be a fatal mistake to allow it to interfere with or in the slightest degree take the place of the social and domestic element, and the personal relationship between the physician and his patient, which tend to mitigate the distress which may be occasioned by the loss of many home comforts and associations, along with the residence amongst strangers.

III. I must pass on now to an important event in the history of the Retreat. I refer to the publication of the "Description of the Retreat,"* written by Samuel Tuke in 1813, and dedicated to his grandfather, William Tuke. Now what had the old York Asylum been doing since the female patient died there in 1791, an interval of 42 years? Why, it had gone from bad to worse. In the Preface to this book the author made an observation which gave great offence to the superintendent, who interpreted it to be a reflection upon that institution. Well, what was this terrible passage? Nothing more than this. "If it" (that is this book) "should be thought to afford satisfactory

* "Containing an account of the Origin and Progress, the Modes of Treatment and a Statement of Cases, with an Elevation and Plans of the Building." Harvey and Darton, London, 1813.

evidence in favour of a more mild system of treatment than has been generally adopted; if it should also prove, which I flatter myself it will, the practicability of introducing such a system into establishments for the insane poor, whose situation has, in general, been too pitiable for words to describe, I shall esteem myself peculiarly happy in this publication." This paragraph *did*, however, cause the greatest offence, and the superintendent of the asylum wrote a warm letter to the newspapers under the name of "Evigilator" in defence of the institution. *Qui s'excuse s'accuse.* From that moment hostilities commenced. York became the scene of an exciting encounter. I have said that Fairfax's battery on Lamel Hill was a symbol of the moral warfare upon which the Retreat entered. I find in the *Yorkshire Chronicle* of September 30th, 1813, a letter from Northallerton, signed by "Viator," which runs thus:—

"It is customary with travellers to call for the papers containing intelligence of the important events which now attract the attention of all the world. After my supper this evening I indulged my usual appetite for news, and on two papers being brought to me, from a sort of instinctive partiality for Yorkshire, I seized the *York Courant*, in preference to a London paper, which was at the same time laid upon the table. The editor's summary account from the late Gazettes pleased me much; I there read: First despatch, 'Forced St. Cyr from a strongly entrenched camp;' second despatch, 'Melancholy fact of Moreau having lost both his legs;' third despatch, ' Important victory over Vandamme;' and fourthly, 'A Gazette containing the numerical account of cannon and prisoners taken in the various actions.'

"My heart was filled with exultation at these glorious achievements of our allies. Nothing less than the humiliation of the Grand Tyrant and the repose of all the world filled my imagination, when casually casting my eye upon a column of the paper parallel to that which contained this gratifying intelligence, I found an account of further hostilities having been carried on by 'storming,' 'boarding,' 'grape or shells, by 'sapping,' 'mining,' 'catamaran,' or 'torpedo.' Now (thought I) for the fall of Dresden! And who is the gallant General that has employed all these means? On looking for the name and the date, I discovered with astonishment that *York* was the scene of these tremendous military operations.

"In a fit of terror and surprise the paper fell from my hand; by an involuntary impulse I rang the bell, and on the waiter entering, anxiously inquired if he had heard that the City of York had been blown into the skies by some insidious revolutionists. With equal surprise, but to my great joy, he answered, 'No, sir, all was well there to-day when the coach left it.' Recovering a little from my confusion, I took courage to examine this article a little more carefully."

The writer tells us that he then found that the article he had read was occasioned by the alarm which one "Evigilator" had taken at a mere description of the Retreat, written by one of the most unwarlike and inoffensive of people.

William Tuke, as vigilant and earnest as he had been in 1791, wrote a letter to the Governors of the York Asylum, in which he says he had the satisfaction of asserting that " kind and conciliating treatment is the best means to promote recovery, as proved in the management of the Retreat, where coercion, though sometimes necessary for feeding the patients and preserving them from injury to themselves or others, is administered in the most gentle manner, and the use of chains is never resorted to." It was not difficult to read between the lines, and the Governors, doubtlessly, did so. And here I cannot avoid pointing out the gratifying contrast, in which no one rejoices more than the present Governors, presented by the well-managed institution of to-day—well-managed for so many years —and that which, unfortunately, became so notorious at the period under review. As a Governor of Bethlem Hospital, I have the corresponding feeling. Nor can I resist the temptation of expressing the pleasure which I feel in the fact that a former superintendent of the York Asylum, Dr. Needham, has been made a Lunacy Commissioner. A better appointment the Lord Chancellor has never made. Writing in the *York Herald* of October 23, 1813, Henry Tuke says of these

Governors:—" Like a modern warrior of declining fame, they claim victory where others consider them defeated. Their self-congratulations will add nothing either to their own credit or that of their cause. The asylum has been wrested from its original design; the poor are in a great measure excluded; and the Institution, it is understood, is committed to the care of a physician and apothecary, without the interference of any committee or visitors in the internal management. Thus, instead of being a public charity, it has become a source of private emolument, and '*hinc illæ lachrymæ*,' Let the Governors of the asylum turn their attention to this important subject, and seriously consider whether they are acting the part of good stewards of the trust reposed in them. It is to them only that the public can look for a reformation, and without their interference all altercation is fruitless."

The question at the bottom of all this controversy was, whether or not the same system of neglect and cruelty, alleged to have been in force in 1791, was still a reality in 1813. As we know, prolonged investigations followed. Concealment was attempted, but fortunately in vain. A Yorkshire magistrate, Godfrey Higgins, of Doncaster, attracted by the fray, and convinced that abuses did exist in the asylum and ought to be exposed, came forward and was of signal service in bringing the engagement to a victorious result. I possess a large number of letters which passed between him and my father at this exciting crisis. A warm friendship was formed between them, based upon their equal indignation at cruelty and wrong. I met the widow of Professor De Morgan, when above 80 years of age, and she told me that she had received from the lips of Mr. Higgins himself a stirring account of his visiting the York asylum one morning, when a remarkable scene occurred. He was assured, on asking the attendant where a certain door in the kitchen led to, that the key could not be found. Mr. Higgins replied that if it was not found he would find a key at the kitchen fireside—the poker. The key was then instantly produced. When the door was opened, this faithful, fearless, and resolute magistrate entered, to find four cells in the most disgraceful and sickening condition. He demanded that he should be taken to see the patients who had slept there the previous night, and was shown no fewer than thirteen women! Comment is needless.

To give a history of this period and the disclosures

which were made, would require a lecture devoted to it; but for our present purpose it is sufficient to record the fact that the Governors of the asylum, with the Archbishop of York in the chair, reinforced by the entrance of a batch of new Governors, eventually passed a series of resolutions which sealed the fate of the old *régime*, and paved the way for a complete reorganization of the management of the institution in 1814. I have met with those who think that the ill-treatment to which the insane were subjected in former days, whether in this asylum or Bethlem, in which I feel as much interest as in the Retreat, should be passed over in silence; I have indeed. But I am strongly of the opinion of Sydney Smith, when he said in anticipation of such a mistaken feeling, and in reference to the abuses in this very asylum at York, that they should be "remembered for ever as the only means of preventing their recurrence."

Now it was undoubtedly the exposure of the condition of the insane in the old York Asylum, followed as it was by suspicions in regard to the state of other asylums, which led to Parliamentary investigation into the abuses which, almost everywhere, existed at that time, and which, happily, forced the Legislature to pass acts for the protection of the insane and for the provision of better institutions. The link between the successful management of the Retreat on new lines and lunacy legislation is not my assertion. It was clearly pointed out by Sydney Smith in 1817, as well as by many others:—"The new Establishment" (he says) "began the great revolution upon this subject," and he adds, "The period is not remote when lunatics were regarded as being insusceptible of mental enjoyment, or of bodily pain, and were accordingly consigned without remorse to prisons under the name of mad-houses, in the confines of which nothing seems to have been considered but how to enclose the victim of insanity in a cell, and to cover his misery from the light of day. But the success of the Retreat demonstrated by experiment that all the apparatus of gloom and confinement is injurious, and the necessity for improvement becoming daily more apparent, a Bill for the better regulation of mad-houses was brought into Parliament by Mr. Rose." It was, sad to say, after great delay and discouragement that really effective Acts of Parliament were passed, and, in this connection, the name of Lord Shaftesbury at once rises to my lips. In the speech which he delivered in the

House of Commons when Lord Ashley, on the occasion of his introduction of the famous Lunacy Bill of 1845, his eulogy of the movement inaugurated here 53 years before, is of the strongest and warmest character. I am sure that we, who know what Lord Shaftesbury has done for the insane, can most fully appreciate the splendid, and, as in the case of the projector of this Establishment, the unremunerated services, which he rendered to this neglected class, and must acknowledge that the work in which he was engaged with such unfailing energy and perseverance was, as he himself regarded it, the necessary supplement to previous reforms, inasmuch as it evoked the strong arm of the law to make adequate provision for the insane and to protect them from harsh treatment. Honour to whom honour is due!

I should like to refer now to one of the most pleasant features of the history of the Retreat, and that is that there has been no international rivalry, and no desire in our own country to detract from the beneficial effect of the courageous step which was taken in this City 100 years ago.

A well-known French physician, the late Dr. Foville, after observing that Pinel was not aware of what had been accomplished at York until 1798, and that on the other hand it was not until 1806 that the news of the enterprise undertaken at the Bicêtre reached the Retreat, generously acknowledges that the philanthropists in Paris and in York alike deserve public recognition for the work of humanity which they contemporaneously accomplished in France and in England, without there being room for raising any question of rivalry or precedency between them.*

And who is there amongst us, as among all British alienists, that does not revere the memory of the illustrious Pinel?

Germany clearly recognized the improved methods of treatment introduced at the Retreat. One day, nearly 60 years ago, there arrived in York a German physician, Maximilian Jacobi, the son of the well-known mental philosopher, the head of a school of metaphysicians contemporary with Goethe, who took a great fancy to the medical son, and expressed his "admiration of his unswerving devotion to his profession." The doctor came to the Retreat, was delighted with what he saw, and stayed some days at York for the purpose of examining on the spot the

* Introduction to " Le Corps et l'Esprit," page xx.

arrangements and management of an Institution with which he had already (in the year 1822) made his countrymen acquainted, by translating into German, the work on the Retreat of which I have already spoken.

I really must read to you the passage in his travels wherein he describes his visit to this City, to which he came by coach from Hull. He says:—" As I approached York I perceived the Retreat through the trees, when looking out to the left of the road, being able to recognize it from the 'Description of the Retreat,' which I had translated, and I rejoiced that I was now able actually to see this memorial of Christian humanity. A letter from my friend, Dr. Zeller, of Winnenthal, secured for me a very friendly reception from Samuel Tuke," who, Dr. Jacobi goes on to say, "introduced me to the superintendent of the Retreat, Thomas Allis, who by his character as well as by his outward man produced a powerful impression, and who possessed special knowledge and dexterity in (comparative) anatomy, as was proved to me by the beautiful preparations to be seen in the new Museum of the Yorkshire Philosophical Society. As Thomas Allis led me through the Retreat I felt at home from the first step, because I had so long been familiar with the plan and arrangements of the building from my translation of Tuke's "Description of the Retreat." I may mention that some time after his return to Germany he sent the latter a work on insanity inscribed, "To his friend, in dear remembrance of the two days spent with him in October, 1834. Siegburg."

Dr. Jacobi became in the course of years the Nestor of German medical psychologists, and, while the Superintendent of the Siegburg Asylum, near Bonn, he in his turn wrote a work on the construction and management of asylums, which my father asked John Kitching to translate, and wrote an Introduction to it of some length. I may add that I visited him at his asylum on the Rhine, when he was in very advanced life, and that he had lost none of his interest in the Retreat, nor was the memory of his visit to York dimmed by age. The whole incident affords a pleasing picture of international reciprocity in the common interests of humanity, and emphasizes the truth of what I am endeavouring to show, that so far from there having been any jealousy on the part of foreign countries, there has been the fullest, warmest, and most generous appreciation of the lead taken a century ago by the Institution whose birth we celebrate to-day. In connection with the visit of Dr. Jacobi to the Retreat, I may mention that another figure in the group to whom he makes a pleasant reference was Dr. Caleb Williams, an honoured name so familiar to us all, and for so many years professionally connected with the Retreat.

The Americans, and notably the very distinguished Dr. Isaac Ray, have been forward to pay their tribute to the influence exerted by " The Retreat," and have acknowledged the direct help they derived in the way of advice from those who were connected with it. I may, perhaps, be allowed to say that I possess the original letter of inquiry from an American to Samuel Tuke respecting the Retreat, and that it was in replying to it, the latter was led to think it might be useful to publish an account of the mode of treatment practised there. This resulted in the work the wide-spread influence of which he little anticipated.

In our own country there has been the same generous feeling in recognizing the position of the Retreat as the pioneer in the amelioration of the condition of the insane. I may specially refer to Dr.

Conolly, for the circumstance which connects his career with the Retreat is exceedingly interesting. I have just spoken of the remarkable influence of the publication of the " Description of the Retreat." But it had another effect no less remarkable, though not so generally known. There was in 1817 in the Edinburgh University a student of medicine of Irish extraction, but born in Lincolnshire, into whose hands there fell this book, and upon whom it produced a powerful and, as it proved, a permanent and far-reaching impression. That student was John Conolly, and in after years, when tracing his past history and the influences which led to his great work, he mentions this circumstance. " Viewing the things which I have described, day after day, and often reflecting upon them, and with deep impression, partly derived from the perusal again and again, even when still a student, of that excellent ' Description of the Retreat near York,' already alluded to, and which I would still urge every student to read and to add to his library, and partly from what I had actually seen at Lincoln a few weeks before commencing my residence at Hanwell, I was not long before I determined that whatever difficulties there might be to encounter, no mechanical restraints should be permitted in the Hanwell Asylum."—(*Medical Times and Gazette*, April 7th, 1860). If that little book of 1813 had done nothing more than inspire Conolly to undertake his work, it would not have been written in vain. Dr. Conolly always took pleasure in attributing to the foundation of the Retreat the reform in the humane treatment of the insane. "The substitution," he writes, "of sympathy for gross unkindness, severity, and stripes; the diversion of the mind from its excitements and griefs by various occupations, and a wise confidence in the patients when they promised to control themselves led to the prevalence of order and neatness, and nearly banished furious mania from this wisely-devised place of recovery."* He spoke of it as "that admirable asylum, the first in Europe, in which every enlightened principle of treatment was carried into effect." I may say that in his declining years I received a letter from him in which he said he loved to dwell upon this theme. I should like to add that we, on the other hand, can and do delight, in the same spirit, to render all honour to the admirable Hanwell physician. My father entertained the highest esteem

* "The Treatment of the Insane without Mechanical Restraints," by Dr. Conolly, page 18.

for him, and in his writings has paid a warm tribute to his " zeal, talents, and integrity." In a letter addressed to myself he writes:—"Lincoln furnished much unhappy evidence in the abuse of non-restraint, and I do greatly rejoice that Dr. Conolly has rescued the great experiment from the failure and miserable reaction which would, I believe, have taken place had it not been for what has really been effected at Hanwell, where all may not be done which meets the eye. I fully believe an excellent system is admirably carried out, and that Dr. Conolly really deserves all the credit which is given to him on the subject. We ought never to have recourse to mechanical restraint at the Retreat, except when it is decidedly the most easy and altogether unexceptional method of coercing the patient; and whenever that is really the case, why should we be subject to a prohibitory law? If the general principle on the subject be fairly carried out, it will, I believe, be found that the infrequency of the exceptions will prove how fully the rule of non-restraint is carried out by us, and this kind of evidence ought to be satisfactory, and will, I think, be so to all reasonable men."

I need hardly say that the writer of this letter raised his earnest protest against the abuse of restraint, and reprobates what in our days it would be a work of supererogation to mention, "those swingings, whirlings, suspensions, half-drowning and other violent expedients by which some physicians have sought to frighten the unhappy subject of insanity into reason, or at least into subjection."*

These observations are necessary in order to understand the position taken in regard to mechanical restraints by those who first undertook the charge of the Retreat. When kindness failed to subdue maniacal excitement, when medical remedies failed to calm, and when there was danger to life or limb of a patient or attendant, then mild forms of personal restraint were reluctantly adopted rather than maintain a prolonged and exasperating conflict between them. It is notorious that at the same period, painful and degrading forms of restraint were employed in many asylums, and even at the Lincoln asylum, so worthily distinguished afterwards for its humane treatment, iron handcuffs weighing 1lb. 5oz. and iron hobbles weighing 3lb. 8oz. were in use until the year 1829.

Having now glanced at the former days of this

* Introduction to Jacobi's "Construction and Management of Hospitals for the Insane," by Samuel Tuke, 1841, p. 35.

Institution, and endeavoured to show the great objects contemplated when it was founded, and having shown that the example it set has exerted a wonderful influence for good by its dual action of exposing abuses, and, most important of all, of showing a more excellent way, I would, in conclusion, emphasize the encouraging record of a century :—

> "Over the roofs of the pioneers
> Gathers the moss of a hundred years ;
> On man and his works has passed the change
> Which needs must be in a century's range."

Happily the moss which has accumulated upon the roof of the building which the pioneers of a new era in the history of the insane erected, has not been an indication of stagnation and desuetude, but rather the venerable reminder of the Past—the original work done under the roof of the dear old Retreat. We gladly recognize that a change has passed over man and his works, such an one as must necessarily be evolved if the law of progress is to be fulfilled. During this period, the civilized world has seen the rise and development of an entirely different system of treatment of the insane, a complete reversal of opinion and practice having taken place. Therefore I hope it has not been uninteresting or unprofitable to recall, as we have done to-day, the history of the movement in the very place of its birth, and where it was cradled with so much thought and fatherly care—the benefits secured by this remarkable reform not being restricted to time or confined to the narrow locality from which it sprang. The progress may seem to have been slow and intermittent, being often impeded by those who ought to have pursued a more enlightened course, but considering the amount of ignorance and neglect, and the time-honoured opinions which had to be exploded, the beneficent change in which all good men rejoice has been effected in a comparatively short period. But here let us be on our guard. There is such a thing as a true and genuinely humane movement against shameful abuses, while on the other hand there is a fussy, intermeddling philanthropy which is as different from the former as the true coin of the realm from the counterfeit. There have been occasions in later times when the pendulum of lunacy reform has swung a little too far, and mischief as well as good has unfortunately been done to the very classes for which such movements (sometimes originated by hysterical agitators) have been ostensibly and ostentatiously promulgated. These popular outcries, when ill-founded and, therefore, unjust, are calculated to have the effect of discrediting attempts at reforms when they are really necessary as they were when the Retreat was instituted. But so it has ever been in the history of all philanthropic movements. There have been uncalled-for and feeble imitations of some great original work, and in the minds of too many people the one is mistaken for the other. A pseudo-humanitarianism has ended in making lunacy legislation vexatious, and calculated to interfere with the prompt care and unfettered treatment of the insane by the asylum physician, whose thoughts are diverted by such means from proper scientific work into that which, as General Sherman would have remarked, carries us back to the day when our mothers taught us the Book of Numbers.

It is a great gratification to me to be able to take any part in this celebration. The Retreat is associated with my earliest recollections. My interest in insanity was inflamed by what I saw and heard respecting the patients here when a boy, and I was mainly influenced in the choice of the medical profession by the desire to be connected with this Institution, and it was within its walls when I was on the medical staff that I was able to find the materials necessary for the preparation, in conjunction with my friend Dr. Bucknill, of the "Manual of Psychological Medicine." These details are, of course, of infinitesimal importance to anyone but myself, and I only mention them as reasons why I myself should feel indebted to the Retreat. My reminiscences before, as well as when I resided here, include very definite memories of the Allises, Dr. Williams, Dr. Belcombe, Dr. Thurnam, the Candlers, and last, but by no means the least worthy, Dr. and Mrs. Kitching, whose sons, I am glad to see, are with us to-day. All had their several and particular merits, their especial characteristics, and if, being human, they had their imperfections, they possessed qualities which in their different ways were of lasting benefit to the Retreat.

It was long after my own connection with it that Dr. Kitching was succeeded by Dr. Baker, to whom it must be a great satisfaction to know that his work here is appreciated, and that he can hand over the management of the Institution to his successor in so satisfactory a condition. It is a satisfaction to those also who have its welfare at heart to know that he will, as Consulting Physician, be still associated with it, and will no doubt initiate Dr. Bedford Pierce into his new office much as William Tuke did George Jepson. I am sure we all desire for Dr. Baker many years of health after his long and faithful services, while for Dr. Pierce we wish a most successful career, honourable to himself and of advantage to the Retreat, animated, as he will be, I hope, by the inspiriting memories associated with its past history.

It ought to be gratifying, I may add, to those connected with the Retreat that the Medico-Psychological Association of Great Britain and Ireland has decided to recognize the importance of this Centenary by holding their Annual Meeting in this city in July, and by making the Medical Superintendent of the Retreat the President on the occasion.

I had intended to offer an apology for having so frequently referred to my own ancestors in connection with its history, but I am assured that this is not necessary. The truth is, I found it to be inevitable if I gave any history at all. It naturally happens that family traditions and papers have given me special facilities for preparing this sketch. I may, indeed, employ, in view of the philanthropic movement we celebrate to-day, the language of the Psalm, as paraphrased in what De Quincey called the Divine Litany of the Church of England :—" O God, we have heard with our ears, and our fathers have declared unto us the noble works Thou didst in their days, and in the old time before them ; " they looked forward in faith and hope ; we can look backward and can witness to-day the fulfilment of their hopes. Those who have listened to their words may well be incited to follow in their footsteps. The lesson is surely writ large and clear in the early history of the Retreat, that not only ought cruelty and neglect in the treatment of the insane to be exposed and denounced, but that those who would reform abuses ought to show a more excellent way. May the course of the future history of this Institution be one of continuous progress, inspired by broad and generous ideas, while conducted on the same humane lines which marked its early life!

CELEBRATIONS OF THE RETREAT CENTENARY.

I.

THE first celebration of the Centenary, held May 6th, 1892, was of a local character, those attending the gathering being chiefly residents in York, or officially connected with the Retreat at the present time or formerly.

We are mainly indebted to the *Yorkshire Herald* for the following notice of this celebration :—

The establishment of the York Retreat is so identified with the commencement of the movement which brought about so beneficent a revolution in the treatment of the insane that its centennial celebration claims an amount of attention which is not limited either to those immediately interested in the Institution or to members of the Society of Friends, with which it is more particularly associated. That the event was regarded with some such feeling was evidenced by the extent of the celebrative gathering which took place at the Retreat last night, and a peculiar interest was imparted to it by the presence of descendants of the Founder of the Institution, and of men and women whose names are revered for their unselfish devotion to its interests, as also by the presentation of several mementoes.

Under the presidency of Mr. James Hack Tuke (Hitchin), at one time the Treasurer of the Retreat, a conference commencing at 5.30 was held in the recreation-room of the asylum, a photographic picture of the company grouped near the front entrance of the main building having been first secured.

The Chairman said they must all feel that their meeting that day to celebrate the 100th anniversary of the founding of the Retreat in 1792 was an occasion of no common or merely local interest, inasmuch as it not only celebrated the founding of the institution, but commemorated the initiation of a movement for the humane treatment and care of the insane which had profoundly benefited that most afflicted and helpless portion of the human race throughout the world, many of whom had hitherto been consigned to " mad houses " where the accepted " treatment " consisted chiefly in imprisonment and chains in filthy cells and other bar-

barities. If the Founder of the Retreat and his friends could be aware of the marvellously beneficial change which has taken place in the past hundred years, would they not join with them in profound thankfulness to the Giver of all good that so great a result had attended their belief in and steadfast following of the Divine law of love and kindness? It was a pleasant thought to him that William, Henry, and Samuel Tuke, representing three generations of his family, were permitted to work together in a cause so dear to each. He believed he owed the distinction of presiding on this occasion to the fact that he was the oldest living descendant of William Tuke, bearing his name, and the only member of his family who could remember to have seen the Founder of the Retreat. Although in the lapse of time the fact had necessarily grown dim, yet he did just remember going when a little over three years of age to take leave of his great-grandfather and receive his dying blessing in 1822. It had always been with sincere pleasure that he had witnessed the various important improvements which had from time to time taken place in the Retreat during the last forty years. None of these had seemed to him of greater importance than the extension of the villa system in addition to the old institutional style of building, a system which he hoped would develop still further in the numerous asylums in this country, in which so many huge and unhomelike structures were to be found. The Chairman then called upon Mrs. Pumphrey, the daughter of a former Superintendent of the Retreat (Mr. Thomas Allis) to read a paper entitled " Recollections of the Retreat as it was Fifty Years Ago."

The paper, not intended for publication, contained a number of interesting incidents and references to former patients, many of them of a droll character.

The Chairman announced the presentation to the Retreat of a number of portraits of those who had been connected with the Institution and had passed away, including several superintendents. The pastels were by H. S. Tuke.

Dr. Robert Baker, the present Medical Superintendent of the Institution, then read a paper on the " Ministry of the Society of Friends to the Insane," in the course of which he said it was good for all of them, whether as communities or individuals, to pause periodically amid the hurry and worry of life's fitful fever and to attempt to climb to some relatively high mountain apart and survey the landmarks of the memorable past. Dr. Baker observed it was nearly a hundred years ago * that there came into the heart of the great alienist-physician Pinel the belief that the insane could be safely, satisfactorily, and humanely cared for without the use of chains. It was one of the most interesting chapters in the history of the treatment of the insane to read how bravely and courageously Pinel acted out his convictions in performing the dangerous duties he undertook. Dr. Baker proceeded to point out that it was a hundred years ago that a similar conviction was reached at York, and it was resolved to introduce a humane system of treatment of the insane. Hence the Retreat, wherein commenced what was long since described as "a government of humanity and consummate skill." Dr. Clouston, when, as President of the Medico-Psychological Association,

* Pinel's nephew, Casimir Pinel, discovered in the registers of Bicêtre that the exact date of his noble inspiration was 1793. "On doit croire que ce fut vers les derniers mois de 1793, et non de 1792, que Pinel se presenta à l'hôtel de ville pour demander l'autorisation à la Commune de faire enlever les chaînes aux aliénés de Bicêtre."— "Lettres de Pinel," 1859. M. Semelaigne, the great-grand-nephew of Pinel, gives the date of his nomination to Bicêtre as August 25, and the day of entering upon his duties there as Sept. 11th, 1793.— "Philippe Pinel et son œuvre." Then followed the like humane deed at the Salpêtrière.

he spoke at York in 1889, described the system of treatment adopted at the Retreat as "the keynote, the example to every succeeding hospital in the country. There was no doubt," he adds, "that York was the very Mecca of the mental physician." Probably most of them were aware that in England there were three distinct classes of asylums : 1st, the vast county asylums ; 2nd, private asylums; 3rd, eighteen hospitals for the care and treatment of the insane. "The Retreat" belonged to this latter class, where all the funds derived from the patients who paid were spent on the patients who could not afford to pay. No doubt many of them were deeply attached to the name of "The Retreat," but it was good for them to remember that it was actually and legally a "Registered Hospital" for the medical treatment of persons in mental ill-health ; and it was good for all of them to think of this famous Institution not so much as an asylum as a Hospital for the cure of those many forms of brain disease which collectively were designated insanity. The great lesson that their ancestors taught in entering on their ministry to the insane was that they ought to regard the insane as human beings in affliction, needing not irons and strait-jackets, but kindness, gentleness, patience, and forbearance. Not only did they recognize the fact that insanity was only a form of ill-health, and not a Satanic possession, but that each special case needed to be ministered to according to its own special character and needs. They would agree that in their recent developments at the Retreat, the Society of Friends had acted wisely and humanely in building several villas in their grounds, and in obtaining Belle Vue House, and Gainsboro' House, the Convalescent Home at Scarborough. By means of these villas a higher and healthier classification of their patients was possible, inevitable annoyances of asylum life were minimized, and the prospects of cure considerably promoted. If they visited those villas they would see that they were made gay with plants and flowers, and that home comforts abounded. Asylum surroundings were conspicuous by their absence. There was yet another ministry to the insane, which the Society of Friends had partially adopted at the Retreat, but which they should at no distant date carry out to a much larger degree than had as yet been attained to, and that was the employment of a gradually increasing number of ladies and gentlemen to tend and to associate with the Retreat patients, so that they might be ministered to by someone specially called to his or her high vocation, and endowed with as many as possible of the attributes of the ministering angels of God. In conclusion, Dr. Baker spoke of his impending retirement, after rather more than twenty consecutive years' residence among the insane, and said that he believed that to be called to minister to the insane was to be called to the highest of all ministries but one.

Mr. John S. Rowntree hoped the result of their meeting together would be to excite renewed interest in the Retreat. He believed that the Retreat, in common with other institutions of the Society of Friends, had suffered some loss of interest from the origination of those great movements which had called their sympathies out of the narrower channel in which they had hitherto flowed into the wider and more national ones. He thought there was great force in the remarks of Dr. Baker respecting the employment and special training of young people for association with the insane.

After an interval for refreshments,

Mr. William Pumphrey submitted a paper, entitled, "The Retreat Hospital for the Insane viewed as a Social and Financial Factor," in the course of which he sketched the various changes which had taken place in the constitution of the Retreat, detailed its mode of working, and gave statistics of its financial position. The original amount of the donations was £30,000. Patients had benefited in consequence of the low rate of charges to the poorer class to the extent of £99,000, and yet the property of the Institution was now valued at £52,000.

Dr. D. Hack Tuke then read his paper on the "Early History of the Retreat, its Objects and Influence."

Dr. Baker moved a vote of thanks to Mr. Tuke for his kindness in presiding.

Mr. Fryer seconded, and Mr. Joseph Rowntree supported the resolution.

The Chairman having responded, the proceedings became of a conversational character, and shortly afterwards terminated.

II.

The second celebration of the Centenary took place in connection with the Annual Meeting of the Medico-Psychological Association, July 21st, 1892, at the Retreat, the Medical Superintendent, Dr. Robert Baker, being President. Dr. Semelaigne, the great-grand-nephew of Pinel, and Dr. Jules Morel, of the Hospice Guislain, Ghent, were among the visitors.

Dr. Yellowlees said that it seemed to him that the Association could not do less than adopt some resolution embodying their appreciation of the benefits conferred upon the insane by the movement which commenced a hundred years ago. He, therefore, proposed the following resolution :—"That the Medico-Psychological Association of Great Britain and Ireland, assembled in its Annual Meeting at the York Retreat in the year of its Centenary, desires to place on record its admiration of the spirit which animated William Tuke and his fellow-workers a hundred years ago, its appreciation of the mighty revolution which they inaugurated, and its thankfulness for the beneficent results which their example has secured in the humane and enlightened treatment of the insane throughout the world." He had thought it desirable to put the Resolution in the plainest words that he could, feeling that language of a fulsome character would be out of place in paying their tribute to such men. They were earnest, God-fearing men, who loved their fellows, and who gave all the kindly help they could to the men who needed it most. They were men actuated by the highest motives, men of sound judgment and wise action, and he wished that all those who had appreciated their motives had emulated their wisdom.

They were no faddists who were carried away by ideas, and still less were they Pharisees who attempted to earn the good opinion of men. William Tuke when he built that Retreat never imagined that he was building a famous name. It seemed to him that they were better able to appreciate the great work that he did in those days by reason of their distance, and they could realize that it was really a revolution. It was something more than dispensing with needless restraint. It was a revolution—a recognition that insanity was a disease, not a doom, and that insane people needed sympathy, kindness, and care instead of the harshness and cruelty which they had hitherto received. The results of their work they, too, could better appreciate. It took a hundred years to tell how a great work would proceed, and they knew now how mighty the change had been. The contrast between the condition of things before the establishment of the Retreat and the condition of things now was the contrast between light and darkness. It was one of the greatest triumphs of humanity and philanthropy that their era had seen.

Dr. Whitcombe, ex-President, cordially seconded the resolution.

Dr. Jules Morel, the President of the Society of

Mental Medicine of Belgium, desired to say that he agreed with every word of it.

The Resolution was very heartily carried by acclamation, upon which Dr. Hack Tuke presented the first copy of the " Dictionary of Psychological Medicine" to the Retreat Library as a Centennial offering.

After partaking of luncheon provided by the Retreat Committee, the members were grouped in front of the Institution and were successfully photographed.

The afternoon meeting was held on the lawn, under the shade of the trees.

Dr. Baker presided, and called upon the Honorary General Secretary, Dr. Fletcher Beach, to read some of the numerous letters received expressing their writers' regret at being unable to be present at the Celebration. These included the following communications:—

From the Commissioners in Lunacy.

Office of Commissioners in Lunacy,
19, Whitehall Place, S.W.,
18th July, 1892.

SIR,

The meeting of your Association at the Retreat at York in this the Centenary year of that Institution affords an opportunity of which the Commissioners in Lunacy desire to avail themselves of expressing their high appreciation of the humane principles of treatment of the insane first practically introduced into this country by its founder, and since constantly applied there.

The value and importance of those principles were fully recognized by the Commissioners' predecessors, the Metropolitan Commissioners in Lunacy, who in their Report for 1844 referred to the Retreat in the following terms:—

"The Retreat at York was established in the year 1792, and introduced a milder system of managing the insane than any then previously practised. This admirable Institution has from its foundation up to the present time steadily preserved the same humane and benevolent method of treating its patients with which it commenced."

The Commissioners are satisfied that these words are equally applicable at the present day.

I am, Sir, your obedient Servant,
G. HAROLD URMSON,
Secretary.

The Secretary of the Medico-Psychological Association.

Letters were also received from the Medical Commissioners, Mr. Cleaton, Dr. Southey, and Dr. Needham, expressing regret at their inability to attend.

Scotch Lunacy Board.
From SIR ARTHUR MITCHELL, K.C.B.
General Board of Lunacy,
Edinburgh, 9th July, 1892.

DEAR SIR,

I beg to thank the Council of the Medico-Psychological Association for their invitation to be present at the annual meeting of the Association to be held in York on the 21st of July, under the presidency of Dr. Baker, in honour of the Centenary of the foundation of the Retreat. I greatly regret that, in consequence of the state of my health, I cannot accept the invitation; but, though not present, I shall join most heartily in the celebration of an event which has proved so great a blessing to the insane of our country and of all countries.

The whole work of my life has been coloured by Samuel Tuke's "Description of the Retreat." It was William Tuke who founded the Retreat, but it was Samuel Tuke who made it known to me, and I think I lift my hat as high to the grandson as to the grandfather. If the " Description of the Retreat " had not been written, I might have been well up in years before I had known much or anything about it. Samuel Tuke's Description spread the story of William Tuke's good deed, and brought imitations everywhere—filled men with the desire to do likewise.

The title of Tuke's work misleads. It is much more than a description of the Retreat. It is a presentation of the principles which should guide men in treating and caring for the insane. It is beautifully written, and I find it still delightful and instructive reading. Our friend Dr. Hack Tuke should be proud of having such ancestors. And proud he is, I doubt not, for he inherits their spirit as well as their name.

I hope you will have a very successful meeting.

Believe me, very faithfully yours,

ARTHUR MITCHELL.

Dr. Fletcher Beach.

From DR. SIBBALD, *a Commissioner in Lunacy for Scotland.*

General Board of Lunacy,

Edinburgh, 11th July, 1892.

DEAR DR. FLETCHER BEACH,

I have to thank the Council of the Medico-Psychological Association very sincerely for their kind invitation to the Annual Meeting to be held at York.

It is with great regret that I find myself unable to avail myself of this invitation, especially on account of the connection of the meeting with the Centenary of the foundation of the Retreat.

I gladly take this opportunity, however, of expressing my hearty concurrence in the intention to do honour to the projector of the

Retreat. No one who is interested in the welfare of the insane can fail to be grateful to William Tuke and his associates and successors in that Institution, where those principles were first carried into operation, upon which the efficient treatment of insanity must always rest.

Had it not been for the Tukes and their fellow workers, one of the most gratifying chapters in the history of British philanthropy might not have been, as it is, a chapter of which we are proud.

With earnest wishes for the success of the meeting, believe me,

Yours very truly,
JOHN SIBBALD.

Dr. Howden, the Medical Superintendent of the Montrose Royal Asylum, wrote a letter regretting his inability to attend.

From the Irish Lunacy Board.
Office of Lunatic Asylums,
Dublin Castle, 19th July, 1892.

DEAR SIR,

Since we cannot attend in person, may we ask you to convey to the members of the Medico-Psychological Association, assembled at York on the 21st July, our warm congratulations on the celebration of the 100th anniversary of the York Retreat, a place ever memorable as the fountain-home of the system of non-restraint in the British Isles, from which the first step was taken to banish the dark ages of cruelty and terror, and to inaugurate a new era in the humane treatment and care of those who, owing to mental defect or perversion, are unable to protect or help themselves.

The Founder of the York Retreat, William Tuke, was like his great compeer, Pinel, one of the truest philanthropists of all time, and to his memory and to his descendants is due a tribute of gratitude from all those interested in the care of the insane in every part of the British Empire, and from no country can it be more heartily offered than from Ireland, where his great work has received such heartfelt sympathy.

A Centenary celebration, which must ever be a landsmark in the study of psychology, should instil in our minds the desire to emulate the great work of the illustrious family, who, discarding old methods and treatment, inaugurated the great work of reforming the madhouses of old, and of freeing the patients from fetters and restraint, and a thousand inhumanities.

     We are, Sir,
      Your obedient servants,
       GEO. PLUNKETT O'FARRELL, M.D.
       E. MAZIERE COURTENAY, M.D.

To Fletcher Beach, Esq., M.B.,
 Hon. General Secretary Medico-Psychological Association.

From DR. LOCKHART ROBERTSON, *Lord Chancellor's Visitor in Lunacy.*

     The Drive, Wimbledon,
      July 10th, 1892.

DEAR SIR,

 I extremely regret that I shall be unable to avail myself of the invitation which the Council of the Medico-Psychological Association have honoured me with for the 21st inst. Had it been a week earlier, when I shall be in York, I should gladly have availed myself of the opportunity you afford me of meeting Dr. Baker and many other of my old friends. But I am due in Edinburgh on the 18th inst., and I have an important professional engagement there on the 20th or 21st which I cannot alter.

     Believe me, sincerely yours,
      C. L. ROBERTSON.

Dr. Fletcher Beach.

From Sir James Crichton Browne, *Lord Chancellor's Visitor in Lunacy.*

Queen Anne's Mansions, St. James's Park, S.W.,
July 6th, 1892.

Dear Dr. Fletcher Beach,

I am much gratified by your courteous note, and sincerely wish it were in my power to avail myself of the invitation which it conveys, for nothing could give me greater pleasure than to meet a group of old friends and colleagues in medico-psychological conclave assembled, on ground, too, hallowed by a century of the calm and persistent pursuit of Humanity in the treatment of the insane. But alas! I have official duties on the date of your meeting which I cannot put aside. Pray express to those assembled at York my regret that I cannot join them, and my unabated sense of fellowship with them in their work, their trials, their aspirations. With kind regards,

Yours very faithfully,
JAMES CRICHTON BROWNE.

Dr. Fletcher Beach, F.R.C.P., etc.

Letters were also received from the President of the Royal College of Physicians, Sir Andrew Clark, Bart., who referred to " the inexpressible benefits conferred on the insane by the Retreat," and the President of the Royal College of Surgeons, Mr. Bryant.

From Jonathan Hutchinson, F.R.C.S., F.R.S., LL.D.,
Ex-President of the Royal College of Surgeons.

15, Cavendish Square, W.
July 16th, 1892.

My Dear Sir,

I much regret that it will not be in my power to be present at the centenary celebration of the Retreat at York. Had it been practicable I should have much liked to avail myself of the invitation with which I have been honoured, to take part in the proceedings. In common with all who are acquainted with the facts I look back with great interest and thankfulness upon the part which was taken by the Founder of the Retreat in bringing about that kindly reformation in the treatment of the insane which has been achieved during the last century. For many years it was almost the only Institution in England in which the poor sufferers from mental disease were received with sympathy, and where the avoidance of all harsh measures was systematically enforced. Nor when the humane principles which it was the first to recognize and to practise had made their way into general acceptance, did this Institution in any way fall behind in the race of progressive improvement. The Retreat has been throughout its whole career, and I believe still is, a model of what may be effected in such establishments by persevering and judicious kindness. In addition to these general considerations I have also personal memories which would have made it a great pleasure to me to take part in the proposed meeting at York. As a pupil of the late Dr. Caleb Williams I long resided in York, and was very frequently, during a period of five years, within the walls of the Retreat. I well remember many of its patients, and with one or two formed friendships which I valued. Under the guidance of the late Dr. Thurnam the foundations of my knowledge of pathological anatomy were laid chiefly in the post-mortem room of the Retreat. I have good reason for remembering the Institution and its officers with warm gratitude, and I wish its Centenary every success.

Believe me, yours truly,
JONATHAN HUTCHINSON.

Dr. Fletcher Beach.

From Dr. FIELDING BLANDFORD, F.R.C.P.

48, Wimpole Street,
20th July, 1892.

DEAR DR. BEACH,

I greatly regret that circumstances prevent my attending the meeting of the Medico-Psychological Association at York. I have a strong feeling of admiration for the work begun at the York Retreat a hundred years ago and carried on since in a way worthy of the founder thereof, and it would have given me great pleasure to have been present on this occasion. With good wishes,

I remain, yours truly,

G. FIELDING BLANDFORD.

Dr. Fletcher Beach.

The President then delivered his Address, in the course of which he said that he had once again to offer them a most hearty welcome to York and to the Retreat. He thanked them most heartily for their courtesy in spontaneously offering to revisit York in celebration of the Retreat Centenary, and also for their goodness in conferring on him just before his retirement from office the high honour of the Presidentship of their Association.

Dr. Baker's Address was mainly devoted to a description of the improvements and additional buildings which have been carried into effect in recent years, more especially referring to the separate villas in the grounds, each carefully designed for the individual treatment of small groups of selected patients. There is the Gentlemen's Lodge, with accommodation for 30 male patients, so planned that it is practically three small independent asylums, each section being fitted with every known appliance for the prompt treatment of each patient. A few years ago the Committee purchased the adjacent estate of Belle Vue House and land for ladies. Further, a house, East Villa, was purchased, with accommodation for three patients. Lastly, there is the West Villa, accommodating from 12 to 15 patients.

Dr. Whitcombe proposed, and Dr. Conolly Norman seconded, a vote of thanks to Dr. Baker for his Address.

A few observations were offered by Dr. Morel and Dr. Semelaigne, appropriate to the occasion.

DINNER.

In the evening the members attending the meeting, together with a number of specially invited guests, were entertained at dinner, Dr. Baker presiding.

The Hon. Secretary, Dr. Fletcher Beach, read additional letters of non-attendance and congratulations to the Committee of the Retreat on its completion of the Centenary :—

From the American Medico-Psychological Association.
Buffalo State Hospital,
Buffalo, N.Y., July 7, 1892.

TO THE PRESIDENT OF THE MEDICO-PSYCHOLOGICAL ASSOCIATION
OF GREAT BRITAIN AND IRELAND.

We take the occasion of the Centennial of the York Retreat, on behalf of the American Medico-Psychological Association (formerly the Association of Medical Superintendents of American Institutions for the Insane), to express the indebtedness of the alienists of America to the York Retreat and to the pioneer work of its Founder in bringing about the improved treatment of the insane. The reform in the treatment of this unfortunate class, inaugurated by the establishment of this Institution, and the principles confirmed by its experience, have gone forth to their beneficent work for successive generations to every land where the English tongue is spoken or English thought dominates public sentiment. The importance of this work has had fresh emphasis during the past ten years in America, where the methods of managing insane patients have been practically revolutionized by discarding mechanical restraint and promoting the employment of every class of insane patients. Many officers of American institutions for the care of the insane felt renewed courage to undertake these reforms after visiting the York Retreat and observing personally what had been accomplished there.

It should be a matter of congratulation to the descendants of William Tuke that the good work which he began one hundred years ago has been increasingly effective year by year since. Kindness, tact, and employment seem very simple means to accomplish such wide-reaching results, but they have proven more effective in the management of the insane than the sterner measures formerly in use. The physicians of America engaged in the treatment of the insane beg to join with the British Medico-Psychological Association in doing honour to the memory of those pioneers in the humane treatment of the insane who bore the name of Tuke.

With great respect, we remain,

J. B. ANDREWS,
President.

HENRY M. HURD,
Secretary.

Dr. Hurd, of the Johns Hopkins University, Baltimore, in forwarding the foregoing, expressed his hope that Dr. Walter Channing, of Boston, then visiting England, would be able to present it to the Retreat meeting on behalf of the American Association, but unfortunately his engagements obliged him to return home before the day of the Celebration.

From Dr. JOHN CURWEN, *Medical Superintendent of the Hospital for the Insane, Warren, Penn., U.S.A.*

Warren, Penn., July 11, 1892.

DEAR SIR,

It gives me great pleasure, as one of the oldest members of the American Medico-Psychological Association, to be able to send a most hearty greeting to the British Medico-Psychological Association assembled in the ancient city of York to commemorate the great event in the history of the care of the insane in England, instituted by William Tuke at the Retreat.

Believing fully in the practice commenced at that time at the Retreat that restraint should only be used as a means of protection to the individual, the effort has been constantly made to minimize its use.

We need to have our thoughts directed more earnestly and intently on a greater variety of diversion and occupation for all the insane, as that seems to be a more direct appeal to the mental structure, while the medical, dietetic, and hygienic treatment build up the physical structure.

The American Medico-Psychological Association expects to celebrate its semi-centennial in 1894, when it is hoped that many members of the British Medico-Psychological Association will be able to meet with us, if they do not feel able to attend the meeting in Chicago in June, 1893.

Very cordially yours,
JOHN CURWEN.

Fletcher Beach, M.D.

From Dr. Stearns, *Medical Superintendent of the Retreat, Hartford, Connecticut.*

Hartford, July 4th, 1892.

My Dear Dr. Hack Tuke,

It would certainly give me great pleasure to be present at the meeting of your Association at York, not only because of my present interest in Old York and its vicinity, but especially that I might present in person the greetings and congratulations of the Hartford Retreat to her Elder Sister on the occasion of her centennial anniversary. It is certainly unusual for a younger sister to congratulate an elder one on the attainment of an advanced age, but when, as in the present case, she has long been the mother of many vigorous children who rise up, not only in all parts of Europe, but also in America, and call her blessed, surely congratulations may be considered in order. On this birthday anniversary of our country, therefore, the Hartford Retreat sends salutations and greetings to the York Retreat, and begs to drink to her health.

May the coming century of her life be characterized by the same high purposes, and crowned with the attainment of even greater successes than those of the past. With best wishes for a good meeting.

I am, most sincerely yours,
H. P. STEARNS.

From Dr. John B. Chapin, *Medical Superintendent of the Pennsylvania Hospital for the Insane.*

Philadelphia, July 6th, 1892.

My Dear Dr. Tuke,

It is a subject of regret that I cannot be one of those who will assemble at York, on the 21st, to recognize in some appropriate way the founding of the Retreat, one hundred years ago. It is not so much the fact that at that period improved accommodation was made for a certain number of afflicted and helpless insane persons, but that the principles which actuated the Founder—William Tuke—should be the leading thought on an occasion like that which calls you together.

It is fitting and becoming that the Medico-Psychological Association of Great Britain should commemorate and honour the Centenary of the establishment of the Retreat by holding its Annual Meeting this year at York. Those engaged in the treatment and care of the insane at this day may well come together to bear testimony to the great advances that have been made during the past hundred years, mainly along the lines originated in the action taken by the founder, that they should recognize the fact that those principles of the humane care of the insane which were then inculcated have been universally confirmed by actual experience, and that the present event may be regarded as a milestone in the great march of humanity by all the English speaking people throughout the world.

At the date of the founding of the York Retreat, the Pennsylvania Hospital was the only established institution for the insane in the United States. This hospital has always been largely under the influence and control of the Society of Friends. Many of our contributors and managers have from time to time visited the Retreat to observe its operations, and to derive from the fountain-head a new inspiration for their own work. I voice the sense of the contributors and managers of this hospital when I ask you to be the medium of conveying to the managers of the Retreat the deep sympathy and interest they have in the auspicious event they are about to celebrate, and our congratulations on the direct and indirect results of one hundred years.

   I remain, dear Sir,
     Sincerely your friend,
       JOHN B. CHAPIN,
     Physician and Medical Superintendent.

Telegram from the Russian Medico-Psychological Association.

St. Petersburg, June 20th.   To Dr. BAKER, The Retreat, York.

The Medico-Psychological Association of St. Petersburg congratulates the York Retreat, from which humane ideas were originally propagated throughout the Universe, and contemplates on the occasion of the Centenary the glorious memory of the celebrated William Tuke.

From PROFESSOR MIERZEJEWSKI, *St. Petersburg, Honorary Member of the Medico-Psychological Association of Great Britain and Ireland.*

MY DEAR CONFRÈRE,

I write to inform you that I exceedingly regret my inability to be present at the meeting of the Association held at York on the occasion of the Centenary of the Retreat, but I beg of you to accept the expression of my most cordial felicitation on the occasion of this fête of humanity, which is unique in character, and is associated with glorious memories.

Yours, etc.,

J. MIERZEJEWSKI.

From PROFESSOR BENEDIKT, *of Vienna.*

July, 1892.

MR. PRESIDENT,

My desire to be present at the meeting of the British Medico-Psychological Association was never greater than this year, and I am very unhappy to be prevented enjoying the honour and pleasure.

You celebrate at York a feast in which every friend of civilization must participate with enthusiasm. You in England have, before all, good reason to be proud of this memorial feast. The English can boast to have taken the lead in a great work in which intelligence, nobility of heart, and energy have an equal share.

The combination of energetic manifestation of individualism, with pronounced common sense, exhibited in the features of William Tuke is characteristic of Englishmen, and this national stamp is evident in the great deed at York.

Accept the expression of his greatest esteem from his respectfully affectionate Socius,

PROF. BENEDIKT.

Telegram from the German Association of Psychological Physicians.

Berlin, July 20, 7.50.

The Association of German Psychologists sends its heartiest greetings to the Centenary Meeting of the Retreat, to the Superintendent, to the family of Tuke, and to the Colleagues present at the meeting.

PROF. JOLLY.
DR. LAEHR.

From Dr. Heinrich Laehr, *of the Schweizer-hof, near Berlin.*

July 14, 1892.

Mental physicians have their eyes at this moment directed to the building where for the first time after a long night in which a bitter fate befel the insane, the morning sun shone on their humane treatment. . . . .

How gladly would I have laid on the day of celebration a laurel-wreath upon the foundation stone of the Retreat, and have expressed my good wishes to the English nation, but alas! I am prevented by illness.

German alienists have always had great sympathy with those of England. We have learnt much from them, and still do so. Our younger colleagues travel there and forward to me as Editor of the "Zeitschrift" most excellent articles, and express themselves with enthusiasm as to what they find in England. . . . .

It is justly observed in the last number of the "Journal of Mental Science" that when Jacobi undertook the management of an asylum in his 50th year he, in the first instance, visited England and found in the Retreat a model, in the spirit of which he conducted Siegburg. Thither we young psychiaters directed our steps in order to acquire a practical knowledge of its teaching. Jacobi also made himself personally acquainted with Samuel Tuke and became his warm friend.

I am convinced that in the collective name of German mental physicians I may convey their hearty congratulations on the celebration of this Centenary. Pray assure the assembled colleagues that when they visit our asylums, when they give us their experience, and when they gladden us by their presence, it is to us also a festival. Accept once more the expression of my friendly respect and the cordial greetings of my colleagues by their friend,

HEINRICH LAEHR.

From Dr. HEINRICH SCHÜLE, *Medical Superintendent of the Illenau Asylum (Baden)*.

July 17th, 1892.

HONOURED COLLEAGUE,

Accept, among other hearty greetings, the expression of Illenau's warmest good wishes for the remarkable secular festival of the greatly renowned Institution at York. May it be granted to the famous Retreat to be true to its honourable history; also to continue to be a blessing to the homestead of noble humanity, the handmaid of science, and to us all an example.

Our Illenau also will on the 27th of September celebrate its Fiftieth year Jubilee. United in aims and endeavours, it reaches forth its hand to its elder sister in good wishes—*ad multos annos*.

In fraternal esteem,

Your devoted Colleague,

Dr. H. SCHÜLE.

Dr. H. Tuke.

From M. MOTET, *Ex-Hon. Sec. Société Médico-Psychologique de Paris.*

Paris, July 12th, 1892.

MONSIEUR LE PRESIDENT—HONOURED COLLEAGUE,

I should have been very glad to accept the gracious proof of your sympathy. My regret in being detained in Paris is so much the greater from the sincere pleasure it would have given me to join in the words which will be uttered on the occasion of a glorious anniversary to celebrate the memory of the originator of the York Retreat.

England and France have had as contemporaries two men with generous hearts, who, breaking with the past, have taken pity on the insane, and been the means of emancipating them from their chains.

There is no room for jealousy between them. They have similarly marched onward in the path which sentiments of humanity have thrown open. From this memorable epoch, with both the French and English, the progress in the treatment of the insane dates. It is the duty of our generation to express our gratitude, after the lapse of a century, to the worthy men to whom we owe so much.

I have pleasure in presenting my hearty salutation in assuring you that I am with you on this solemn occasion, and in conveying to you the expression of my respectful sympathy.

I am, Mr. President and honoured Colleague,

Your very devoted,

A. MOTET,

From DR. COWAN, *Netherlands Medico-Psychological Association, Dordrecht, Holland.*

Dordrecht, June 28th, 1892.

GENTLEMEN,

At the last meeting of the Medico-Psychological Association of the Netherlands, on June 22nd, 1892, a Resolution was unanimously passed to congratulate you on the Centenary of the Retreat at York, and to express a hope that a happy retrospect may be yours.

Need we add, gentlemen, that we take part in your rejoicings, and that we sincerely hope the good example set in 1792 may act as a salutary example to all the world, and that the time may come when an *asylum* will be thought of only as a *Retreat* for mental sufferers.

We send you our fraternal greetings, and add the wish that both the

British and the Netherland Societies may long continue in peaceful strife to relieve the sufferings of the insane.

The Medico-Psychological Association of the Netherlands.

Dr. F. COWAN,
President.
Dr. POMPE,
Secretary.

From Switzerland a sympathetic letter was received from Dr. Wilhelm von Speyr, Medical Superintendent of the Waldau Asylum, near Berne.

Speeches were delivered by Dr. Clouston, the City Sheriff (on behalf of the Lord Mayor of York) and Mr. Joseph Rowntree, the Chairman of the Retreat Committee, who proposed the "Medico-Psychological Association," coupling with it the name of Dr. Baker. He thought that the occasion of the Centenary of the York Retreat might be made the starting point of another forward movement. The time of gloomy and forbidding buildings for the insane had passed away, and they had palatial edifices with corridors decorated by Italian artists, and rooms furnished according to the latest teachings of the gospel of æstheticism, but it appeared to him that the Association might be of very great service in creating public opinion on the question of the conditions favourable for the treatment of insanity. If any of them were ever to suffer from that great affliction, he thought there would be something which they would desire more than beautiful rooms, and that would be that they should have companionship and sympathy from men of their own plane of thought and education. Within the lifetime of everyone in that room Miss Nightingale had been able with her wonderful enthusiasm to draw from the educated classes a contingent of ladies willing to enter upon the life of a hospital nurse, and in thinking about that meeting of the Association it occurred to him that probably there might be a possibility that in many of the asylums they should train a body of cultivated attendants willing for a term of years to be the companions of those who were afflicted with insanity.

The President, in responding, said they must feel deeply obliged to Mr. Rowntree for the way in which he had spoken of the work of their Association.

They all felt deep admiration for Tuke, and for Pinel who amidst the throes of the great revolution inaugurated humane movements such as that, the Centenary of which they were now celebrating.

Dr. Yellowlees eloquently proposed the next toast,

"The Dictionary of Psychological Medicine," as fittingly placed on the shrine of the memory of the author's ancestors in their silent presence on the occasion of the Retreat Centenary.

Dr. Tuke expressed his acknowledgments and his unabated interest in an Institution in which he resided many years ago. Over the entrance of a Buddhist Temple in Japan there was an inscription " Stranger, " whosoever thou art, and whatsoever be thy creed, " when thou enterest this sanctuary, remember that the " ground on which thou treadest is hallowed by the " worship of ages," and if an inscription were placed over the entrance to the Retreat, he would suggest this paraphrase :—" Stranger, whosoever thou art, and whatsoever be thy creed, when thou enterest this Hospital, remember that the ground on which thou treadest has been hallowed by a noble deed, and by the humane work of a century." He concluded by proposing the " Health of Dr. Semelaigne," who had come from Paris to be present at this Centenary. He was not only the son of a distinguished alienist in Paris, but was the great-grand-nephew of the illustrious Pinel. They all appreciated the feeling which brought him to York, and the testimony which he bore to the work which the Retreat had performed. With regard to Pinel, there had never been a nobler, never a more humane man in all France. The more he (Dr. Tuke) studied his character, the more he admired him. Therefore it was most fitting that they should on this occasion receive Dr. Semelaigne with the greatest cordiality.

Dr. Semelaigne responded in suitable terms, and observed that two men in France and England, without knowing anything of each other, resolved on each side of the Channel to introduce a humane treatment of the insane. At that moment the two nations were enemies, now they were friends, and the book of wars was closed for ever. As the great-grand-nephew of Philippe Pinel, he was proud to sit amongst them to celebrate the name of William Tuke. He should never forget his journey to York, where he was allowed to see that the two great sister nations had become so friendly and united— England and France, as also two great philanthropic names—Tuke and Pinel.

Dr. Urquhart having proposed "The Visitors," coupling with the toast the names of Mr. W. Hargrove, of the *Yorkshire Herald,* and Dr. Jules Morel, who responded, the proceedings were brought to a close.

The Annual Meeting of the Association and the Centenary of The Retreat, York.

The following leader on the event appeared in the "British Medical Journal," August 6th, 1892:—

"The British Medico-Psychological Association held its annual meeting this year in the city of York, to mark its sense of the benefits conferred upon the insane by the foundation of the Retreat in the midsummer of 1792. Similar Associations in the United States, Russia, Austria, Germany, France, Belgium, Holland, and other countries recognized the interest and importance of the event thus commemorated by sending their greetings. No national rivalries appear to have chilled the expression of the most cordial felicitations on the occasion, and the last number of the 'Journal of Mental Science' contains ample evidence of this generous sympathy in former days. The same Journal contains materials which enable us to appreciate the motives which led to the building of an institution destined to exert so remarkable an influence in reforming the treatment of lunatics in this country.

"Considerable dissatisfaction had been felt for several years prior to 1792 in the management of a Lunatic Hospital at York, established in 1776 by public subscription. In 1791 a lady patient died. Her friends had come from a distance during her illness to see her, but their wish to do so was denied. The event was shrouded in mystery, and suspicions already aroused as to the treatment of the inmates were intensified. A citizen of York known for his philanthropy, and a member of the Society of Friends, took the affair to heart, and proposed the establishment of a new asylum, where the patients should be treated with kindness, and where the feelings of their friends should be consulted. William Tuke could not possibly at that time have a perfect conception of the needs of the insane as we now recognize them, but he broke with the past, and started upon an untrodden path. His merit lies not in writing fine words, but in doing the right thing. Little by little the idea grew and formulated itself, so to speak, in a great work of benevolence and intelligent skill, the outcome of common sense and philanthropy.

"Perhaps, after all, it was an advantage that he had no knowledge of medical custom or theory, for at that period the profession did not shine in its treatment of insanity. In fact, mental medicine was at its lowest ebb, and was summed up in the well-known epigram on Lettsom. Tuke's proposition, coldly received at first, was eventually carried into effect; but for this purpose liberal donations from his co-religionists as well as himself became necessary. He ensured success by residing in and directing the house, and subsequently by obtaining the services of an excellent man, Jepson, possessed of medical knowledge although unqualified, who cordially helped him to carry out his plans. It is evident that a resolute will, strong sense of duty, pity and good sense were essential, and with these qualities the projector of the institution was in a large measure endowed; but more than this, he not only knew where to find his tools, but how to use them.

"We have been at some pains to discover what manner of man he was, and the portrait accompanying the article referred to appears to justify the description given of him in an obituary notice. 'In person, William Tuke hardly reached the middle size, but was erect, portly, and of a firm step. He had a noble forehead, an eagle eye, a commanding voice, and his mien was dignified and patriarchal.' His evidence before the Select Committee of the House of Commons presents a striking picture of the treatment introduced at the Retreat, although evidently not reported *in extenso*. It is satisfactory to know that after more than a quarter of a century's devotion to the welfare of the institution he did not pass away without knowing that the reform in lunacy was progressing, and gave promise of further extension and utility. That he impressed his mark upon his age is proved not only by the quoted testimonies to the contrast presented by the management of the Retreat to that of contemporary institutions, but by the action taken by Parliament in probing the festering wound to the bottom, and initiating lunacy legislation, which by slow yet sure degrees led to enactments made to protect the lunatic and to provide accommodation in asylums which are now the pride of England. With regard to mechanical restraint, its abolition is stated to rest, not with the Retreat, but with Gardiner Hill, Charlesworth, and with Conolly, who attributed his remarkable career in this direction mainly to the Retreat, and observes that, 'although, certainly, restraint was not altogether abolished at that establishment, it undoubtedly began the new system of treatment in this country, and the restraints resorted to were of the mildest kind.' To him the article in the 'Journal of Mental Science' pays a glowing tribute of praise for the ultimate developments of lunacy reform. Now that the battle of humanity has been fought, and the combatants have gone to their rest, their respective share in the work can be and is judged with calm impartiality, and their respective merits justly recognized. This remark applies to those who laboured in France as well as in our own country, and at the dinner of the Medico-Psychological Association at York, a collateral descendant of Pinel was present to do honour to the Retreat on attaining its Centenary, while this physician's health was fittingly proposed by Dr. Hack Tuke, who ungrudgingly paid a warm tribute to the meritorious act of Pinel in the dark days of the French Revolution, in knocking off the cruel fetters of the insane at the Bicêtre. That there should be such a recognition of noble reforms initiated so long ago in the two countries is unmistakable evidence on the one hand of the profound impression they produced, and on the other, of the cordial relations which exist between the alienists of France and England. As we have intimated, no trace of jealousy or rivalry appears in this very pleasing episode. Would that the same happy feeling of international goodwill characterized all the victories of good over evil, and knowledge over ignorance, at home and abroad!"

www.ingramcontent.com/pod-product-compliance
Lightning Source LLC
Chambersburg PA
CBHW020154170426
43199CB00010B/1039